Incredible Insects

by Harriet Loy
Illustrated by
Natalya Karpova

BLASTOFF!
MISSIONS

 BELLWETHER MEDIA
MINNEAPOLIS, MN

Blastoff! Missions takes you on a learning adventure! Colorful illustrations and exciting narratives highlight cool facts about our world and beyond. Read the mission goals and follow the narrative to gain knowledge, build reading skills, and have fun!

Traditional Nonfiction

Narrative Nonfiction

Blastoff! Universe

MISSION GOALS ■ ■ ■ ■ ■ ■

> FIND YOUR SIGHT WORDS IN THE BOOK.

> IDENTIFY THE TRAITS OF INSECTS.

> THINK OF QUESTIONS TO ASK WHILE YOU READ.

This edition first published in 2023 by Bellwether Media, Inc.

No part of this publication may be reproduced in whole or in part without written permission of the publisher. For information regarding permission, write to Bellwether Media, Inc., Attention: Permissions Department, 6012 Blue Circle Drive, Minnetonka, MN 55343.

Library of Congress Cataloging-in-Publication Data

Names: Loy, Harriet, author.
Title: Incredible insects / by Harriet Loy.
Description: Minneapolis, MN : Bellwether Media, Inc., 2023. | Series: Blastoff! Missions. Amazing animal classes | Includes bibliographical references and index. | Audience: Ages 5-8 | Audience: Grades 2-3 | Summary: "Vibrant illustrations accompany information about insects. The narrative nonfiction text is intended for students in kindergarten through third grade"-- Provided by publisher.
Identifiers: LCCN 2022020238 (print) | LCCN 2022020239 (ebook) | ISBN 9781644876503 (library binding) | ISBN 9781648348341 (paperback) | ISBN 9781648346965 (ebook)
Subjects: LCSH: Insects--Juvenile literature.
Classification: LCC QL467.2 .L695 2023 (print) | LCC QL467.2 (ebook) | DDC 595.7--dc23/eng/20220510
LC record available at https://lccn.loc.gov/2022020238
LC ebook record available at https://lccn.loc.gov/2022020239

Editor: Christina Leaf   Designer: Andrea Schneider

Printed in the United States of America, North Mankato, MN.

This is **Blastoff Jimmy**! He is here to help you on your mission and share fun facts along the way!

Table of Contents

Right in Front of You!

Look around you. Chances are there is an insect nearby! These animals have three body **segments**, six legs, and an **exoskeleton**.

Insects come in many
different forms.
Let's go find some!

Dart and Buzz

green darner
dragonflies

The **prairies** of North America
are filled with insects!
Green darner dragonflies dart
over streams to catch food.

Here comes a honeybee! This fuzzy insect is carrying **pollen**. It will make honey back at the **hive**.

honeybee

pollen

▶ **JIMMY SAYS** ◀
Honeybees only sting when they are afraid. They will not attack if they are left alone.

glasswing butterfly

Let's move to the Amazon **Rain Forest** in South America. See the glasswing butterfly fluttering above us? Clear wings help it hide!

Watch out for bullet ants. Their **sting** is the most painful of any insect! Ouch!

bullet ants

Pinching and Jumping

earwig

pincers

Our next stop is Europe! Here, an earwig crawls out from under a rock. It grabs food with its **pincers**.

Playing with Poop

dung beetle

Put on your sunblock. We are in Africa! Look at those dung beetles rolling balls of poop. The poop will be their dinner. Yuck!

These tall mounds are home to a **colony** of termites. These small insects use dirt, spit, and poop to build their homes. Worker termites keep the mounds strong!

mound

JIMMY SAYS

Termites are one kind of insect that have a queen. She lays all of the colony's eggs!

termites

giant hornet

Keep an eye on that giant hornet as we enter this Asian jungle. It has a painful sting!

We almost passed by an orchid mantis! This pretty insect uses **camouflage** to hide.

orchid mantis

Hercules moth

Here we are in Australia! Can you find the hidden Hercules moth? It is the largest moth in the world! There it is on the tree!

JIMMY SAYS

Hercules moths do not have mouths. They cannot eat!

19

We saw many kinds of insects
on our trip. But there are
millions more out in the world!
Which ones can you find?

Insect Facts

three body segments

exoskeleton

six legs

Glossary

camouflage–a way of using colors or patterns to blend in with surroundings

colony–a group of termites that live together

exoskeleton–the hard outer covering of an insect; exoskeletons protect and support insect bodies.

hive–a place where honeybees live

pincers–parts of an insect's body used to grab

pollen–a yellow dust inside flowers and other plants

prairies–lands with tall grasses, flowers, and few trees

rain forest–a thick, green forest that receives a lot of rain

segments–parts; insect bodies have three segments.

sting–the act of piercing with a stinger

To Learn More

AT THE LIBRARY

Messner, Kate. *Insect Superpowers*. San Francisco, Calif.: Chronicle Books, 2019.

Sabelko, Rebecca. *Bee*. Minneapolis, Minn.: Bellwether Media, 2021.

Thomas, Isabel. *One Million Insects*. London, U.K.: Welbeck Publishing, 2021.

ON THE WEB

FACTSURFER

Factsurfer.com gives you a safe, fun way to find more information.

1. Go to www.factsurfer.com.

2. Enter "incredible insects" into the search box and click 🔍.

3. Select your book cover to see a list of related content.

BEYOND THE MISSION

> MAKE AN INSECT FROM THINGS AROUND YOUR HOME. WHAT MATERIALS WILL YOU USE? WHY?

> WHAT INSECT DO YOU THINK IS THE MOST INTERESTING? WHY?

> CAN YOU NAME AN INSECT THAT WAS NOT MENTIONED IN THIS BOOK?

Index